Kindergarten Science Volume 1

© 2013 Todd Deluca
OnBoard Academics, Inc
Newburyport, MA 01950

800-596-3175
www.onboardacademics.com

Table of Contents

Seasonal Behavior

We change our behavior (things like the clothes we wear, the food we eat, and the activities we do), according to the seasons. Plants and animals also change their behavior according to the seasons.

Write the season under each picture.

Animals and plants adopt three main types of seasonal behaviors: they **adapt** (do something differently), they **migrate** (move someplace else and then come back), or they **hibernate** (sleep for a long time).

Adapt

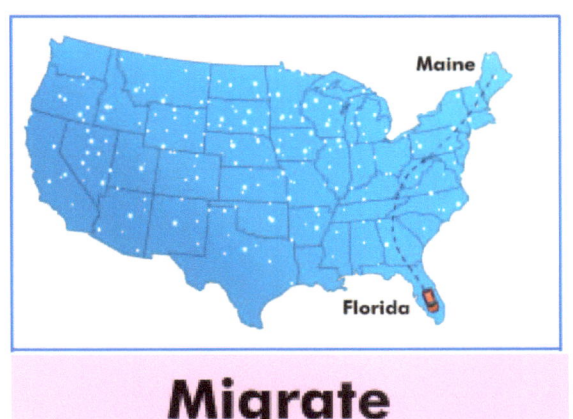

Migrate

Hibernation

Label each picture with the correct seasonal behavior.

Review these three animals seasonal behaviors. What type of seasonal behavior do these three activities represent? _____

In the spring and summer, foxes eat lots of insects, berries, and grasses.

In fall and winter, foxes eat more rodents like mice, voles and squirrels. This is because many plants and insects are either inactive or dead and so foxes have to change what they eat.

When acorns and other nuts are available in summer, squirrels bury them in the ground. Why do they do this?

Trees will not be making acorns and other nuts in winter and so squirrels plan ahead and store food to eat later.

As winter arrives, the fur on the snowshoe hare changes from short and brown to thick and white. Why does this happen?

The snowshoe hare changes its fur to white and thick in winter for better camouflage and to be warmer.

Plants also adapt to seasonal changes. Label each picture with the correct season.

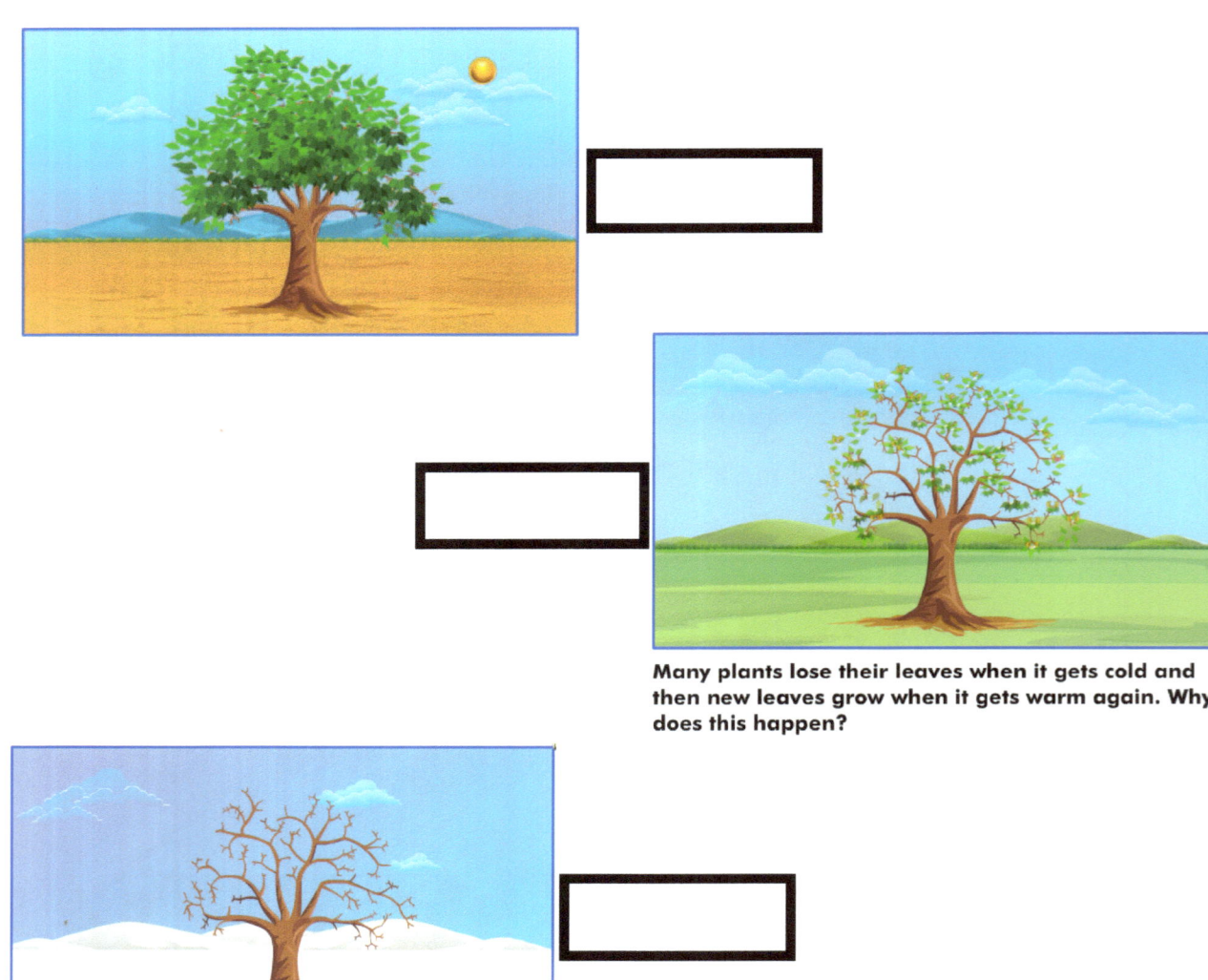

Many plants lose their leaves when it gets cold and then new leaves grow when it gets warm again. Why does this happen?

Plants use leaves to make food energy from sunlight. Many plants and trees stop growing during the fall and winter and so there is less need for leaves. Also, plants lose water through their leaves and so losing leaves helps them to conserve water.

Why do you think the Canadian goose often spends summer in Canada and the Northern U.S. and then spends winter in the Southern U.S. and Mexico?

_____

_____

_____

Groundhogs (also called woodchucks) normally go into their underground burrow in October and then hibernate until March or April. Why do they do this?

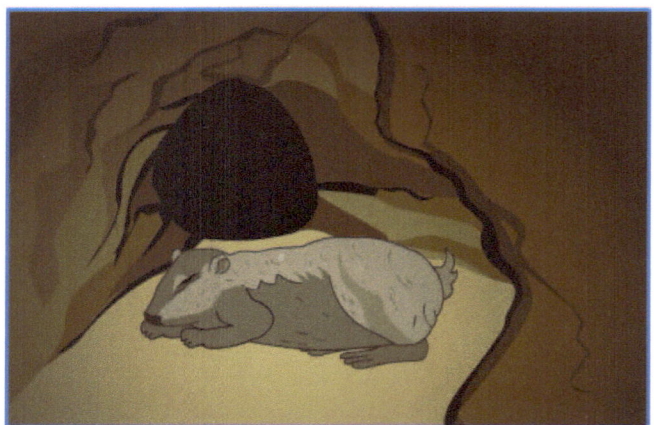

Groundhogs hibernate because the plants and insects they eat are not available in winter, but before a groundhog hibernates, it eats a *lot* of food. When it is hibernating, its body slows down to conserve energy.

Read about each animal and label their activity as adaptation, migration or hibernation.

In summer deer eat acorns and other nuts, green plants and fruit. In the winter, they mostly eat bark because that's whats available.

Monarch butterflies fly over 1,000 miles to avoid cold winters.

In November rattlesnakes head for their favorite den and don't emerge until April.

When it's cold beavers eat tree bark. When its warmer Beavers eat aquatic plants and the leaves and twigs of many trees.

Read about each animal and label their activity as adaptation, migration or hibernation.

Chipmunks sleep through the winter in order to conserve energy and emerge when food is available in the Spring.

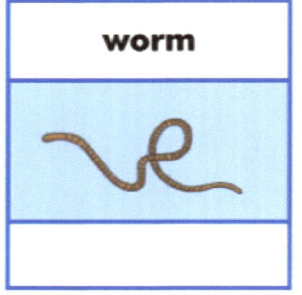

When it gets cold and the soil freezes, worms move several feet deeper into the soil in order to stay warm.

Whales swim thousands of miles to follow the food that they eat.

Seasonal Behavior Quiz
Fill in the blanks

Animals change their _____ according to the season. Some animals travel to a new location in order to find food and to keep warm; this is called _____. Other animals go into a deep _____ (sometimes for many months) in order to conserve energy; this is called _____. Finally, some animals _____ what they do in different seasons. For example, they might store _____ to eat later, change how they _____ to be more camouflaged, or eat different types of food. Plants also change during the seasons. Many plants drop their leaves to _____ water.

food   sleep   conserve   eat   migration

nest   adapt   behavior   look   hibernation

Food Chain

What is happening here?

_____

_____

_____

_____

_____

_____

_____

_____

_____

_____

_____

_____

Food energy is being transferred from one living thing to another.

A food chain shows the order of energy flow through organisms that live in the same environment which we call a biome.

All food chains start with the sun which provides the energy that plants use to make their own food. In this food chain the plants, the grass, is then eaten by a grasshopper. This is how the grasshopper gets energy. The grasshopper is then eaten by a frog. The frog is then eaten by a snake. Finally, the snake is eaten by a hawk. This order of feeding relationship or energy flow is called a food chain.

Roles in a Food Chain

It's easy to identify the role of a consumer in a food chain if you remember that food chains always begin with a **producer**, and that **primary** means first, **secondary** means second, **tertiary** means third, and **quaternary** means fourth.

The living thing at the start of the food chain that makes		**producer**
Living things that eat other living things are called consumers. In this food chain the grasshopper is a primary consumer meaning it's the first consumer in the food chain		**primary consumer**
The frog eats the grasshopper and is the secondary consumer. This means it is the second		**secondary consumer**
Animals that feed on secondary consumers, in this case the snake, are called tertiary consumers.		**tertiary consumer**
Animals that feed on tertiary consumers, in this case the hawk, are called quaternary consumers. Quaternary means fourth.		**quaternary consumer**

Identify the role of each living thing in this food chain.

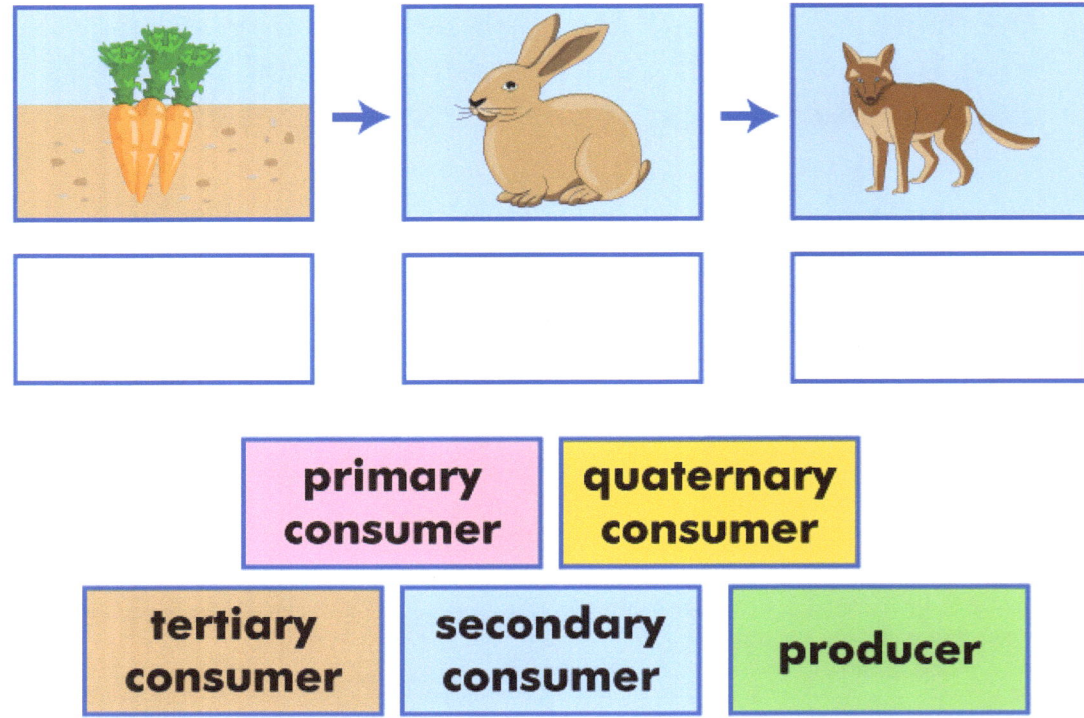

primary consumer | quaternary consumer

tertiary consumer | secondary consumer | producer

A food chain always starts with a producer, but the number of consumers can vary. This food chain shows only two consumers. Food chains rarely go above quaternary consumers (the level above tertiary).

Create a grassland biome food chain with the pictures and labels below.

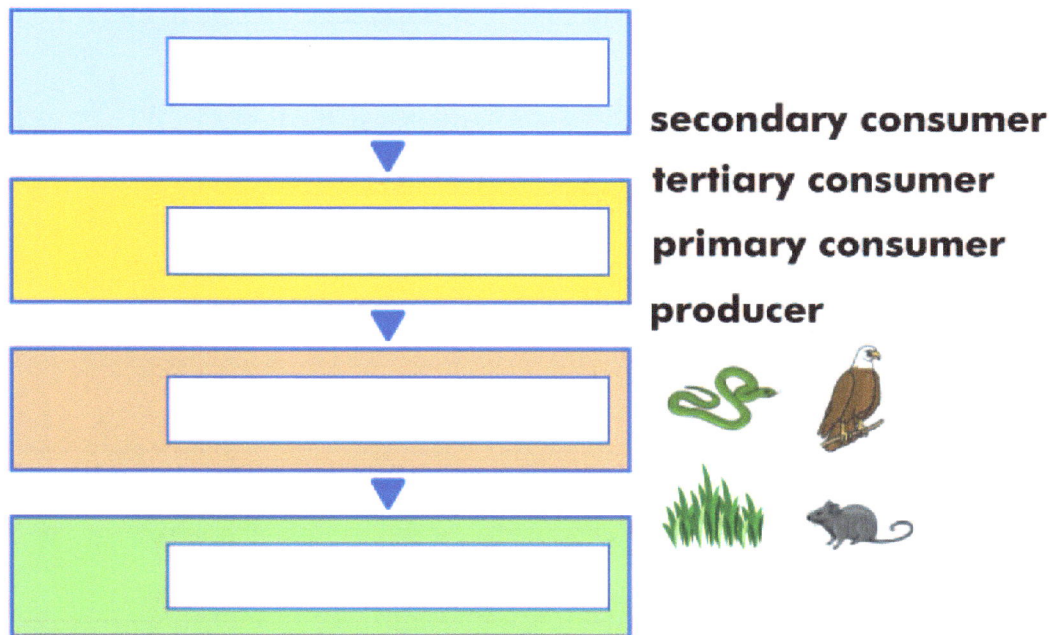

secondary consumer

tertiary consumer

primary consumer

producer

Create an ocean biome food chain using the information below.

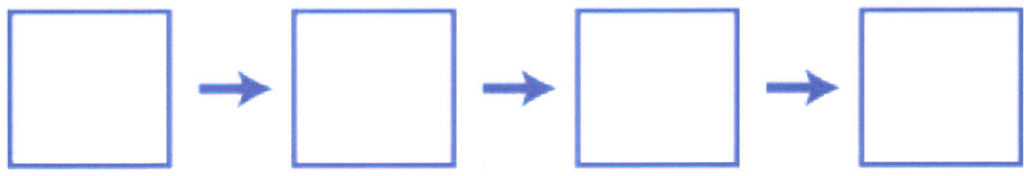

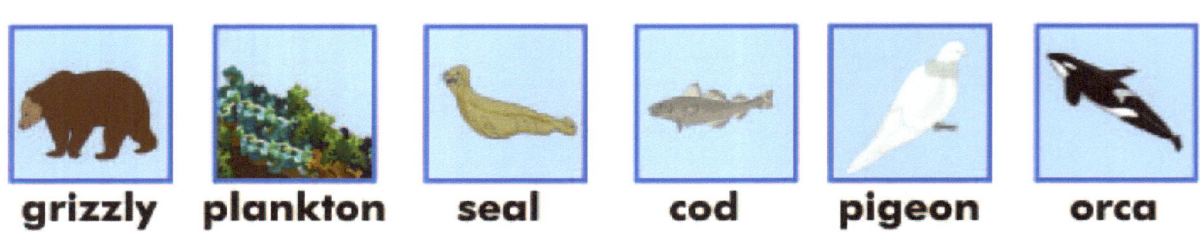

grizzly   plankton   seal   cod   pigeon   orca

Food chains exist between organisms that live in the same biome. The grizzly bear and pigeon do not live in the same biome as the rest of these organisms and so would not be a part of this food chain.

Create a pond biome food chain.

Draw the food chain steps based on the illustration of the pond.

In most biomes, multiple food chains overlap and connect with each other. A network of all the food chains in a biome is called a food web.

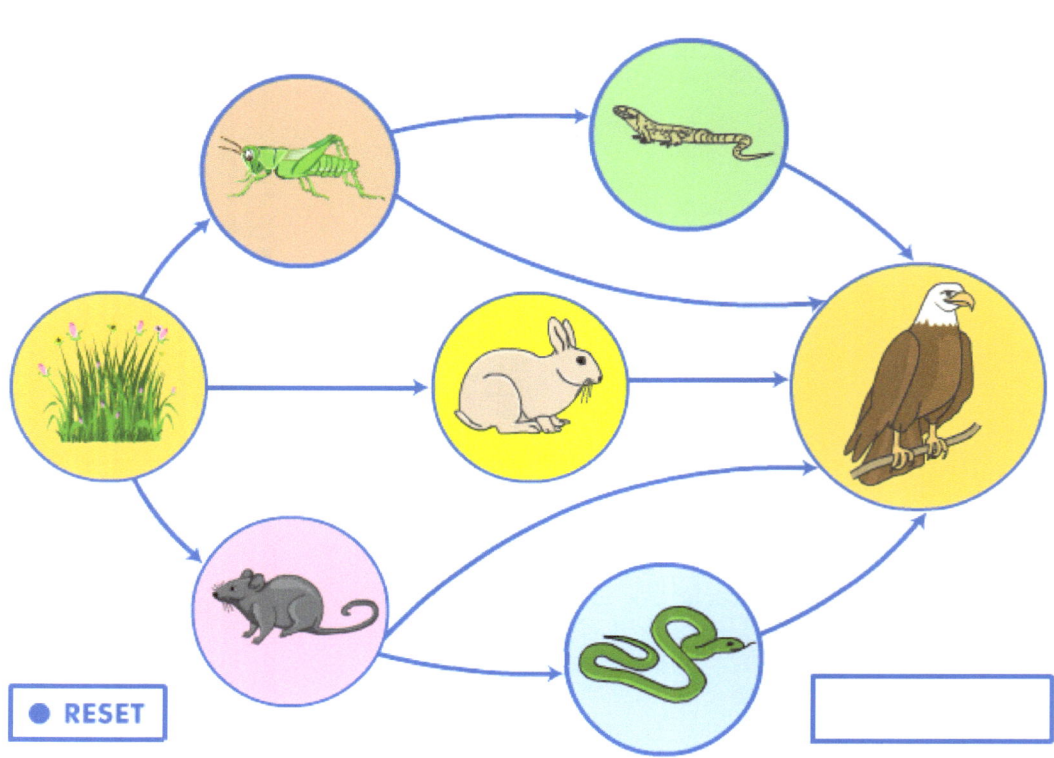

● RESET

Food Web

In a food web, some living things can assume more than one role depending on the food chain within the web. This is why the eagle can be considered a secondary or tertiary consumer.

Based on the food web illustration below, place a check mark in the box to identify the roles of these organisms.

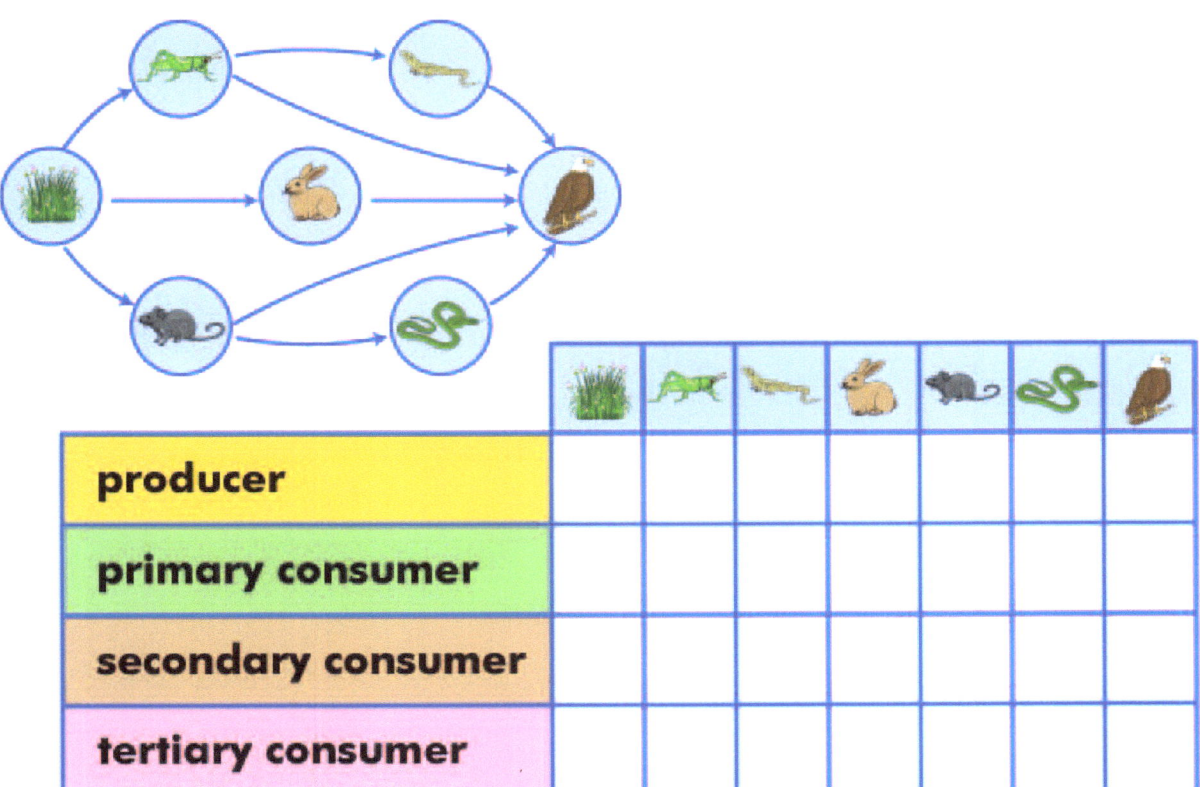

	🌱	🦗	🦎	🐰	🐭	🐍	🦅
producer							
primary consumer							
secondary consumer							
tertiary consumer							

Build a food web by connecting these organisms.

Scientists sometimes use other names to describe the roles of living things in a biome. An herbivore is a living thing that eats only plants. A carnivore eats only animals. Omnivores, such as humans, eat both plants and animals. A predator is an animal that eats another animal (its prey).

Use the label is the colored boxes to identify the different organisms.

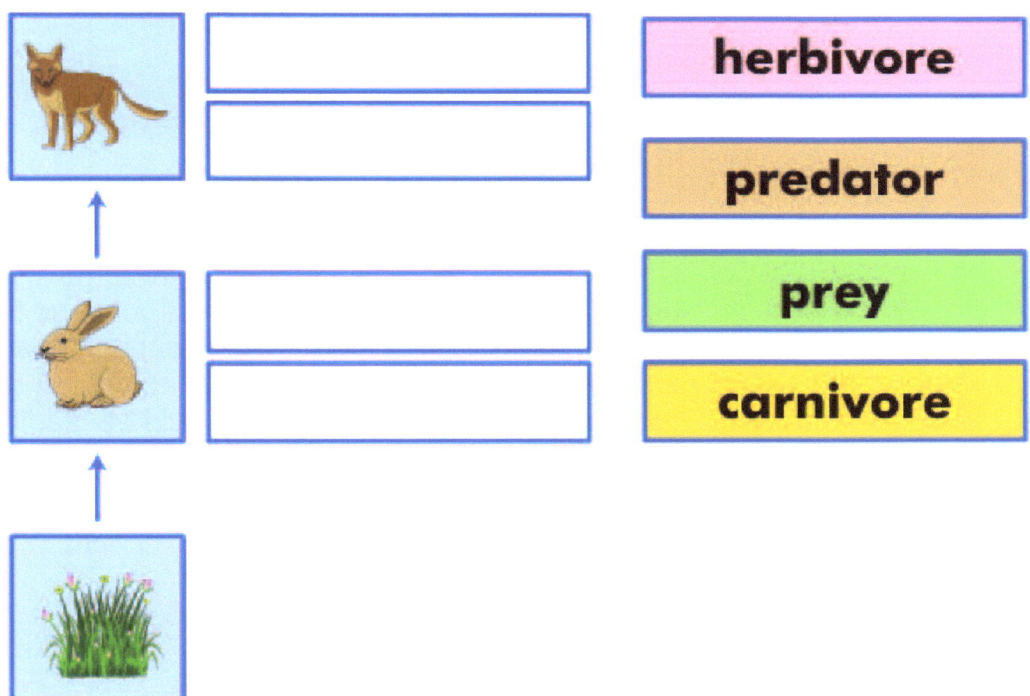

The Importance of Balance in the Food Chain

Within the biome producers are the greatest in numbers because they are the basis of the food chain or food web. Similarly there are more primary consumers than secondary consumers and more secondary consumers than tertiary consumers. This is because each level of the food chain depends on the level below if for its energy. That's why the number of organisms in each level of a food chain look like a pyramid.

If a change occurs at any level of the food chain it can have a big effect on all other levels of the food chain. In our lake biome, imagine what would happen if the number of small fish were significantly reduced due to disease. The large fish would go hungry and would start to die out. This would then impact the bears because they wouldn't have enough food to eat.

Now lets see what would happen if we removed the bears for the food chain. The number of large fish would increase since there weren't enough bears to eat them. This would decrease the number of small fish because so many large fish would be feeding on the small fish.

> **The balance in a food chain is very delicate and a change at one level in the chain can affect the other levels... sometimes in disastrous ways.**

Food Chain Quiz

1. Place the follow in order to form a food chain; snake, grass, chick, caterpillar.

_____ _____ _____ _____

2. In a biological community, an organism that east the primary producer is what? Circle your answer. Producer Decomposer Consumer None of these

3. A food chain always starts with plant life. True or False

4. What is a food web? Place a check mark next to the correct answer.
   a.   A network of food chains
   b. Provides a connection between trophic levels and a biological community
   c. Both a and b

5. Prey kill and eat other animals. True or false?

6. Organisms that eat both plants and animals are called what? Circle your answer.

   a.herbivores   b. decomposers c. omnivores d. detrivos

Predators and Prey

What is this owl most like to eat? Draw your answer next to the owl.

Predators and Prey

Predators are animals that hunt and eat other animals for food. There are many different predators throughout the world and they come in all shapes and sizes.

Prey are the animals that predators eat. This means that wherever there are predators there are prey. Like predators, prey come in all shapes and sizes.

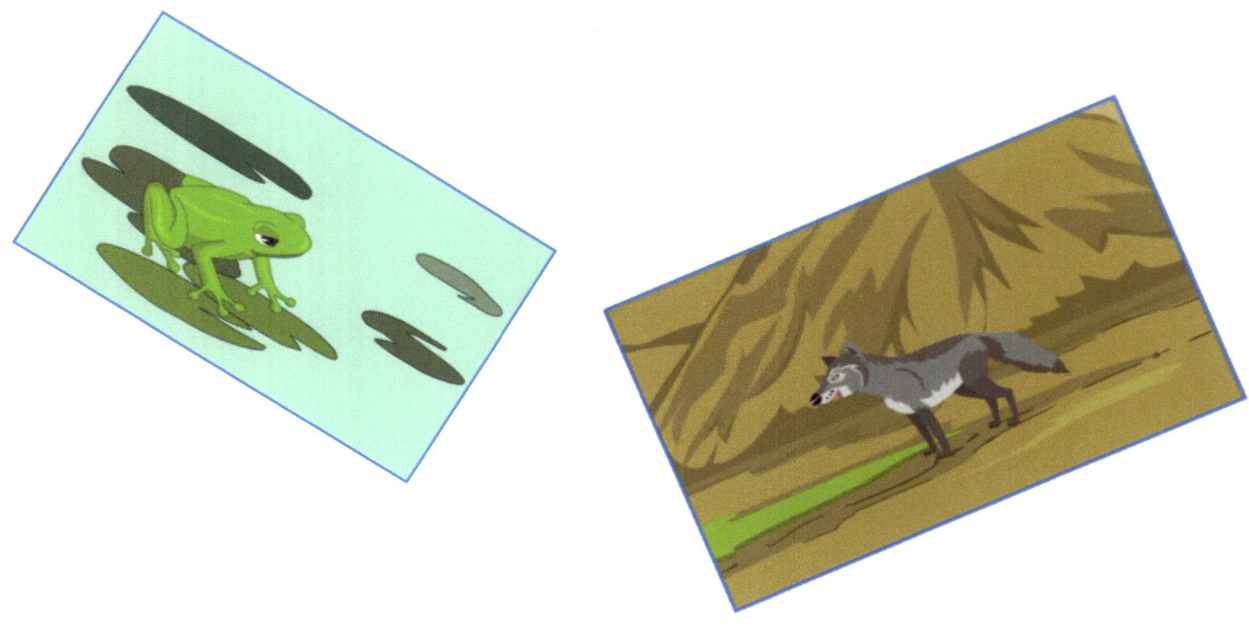

Match each predator with its most likely prey.

Animals often have many prey or many predators.

| Many Prey | Many Predators |

Many animals are both predators and prey.

Within a food chain, many animals have dual role as both predators and prey.

For example, in this food chain the frog is the predator because it eats the grasshopper.

However, the frog becomes the prey as soon as the snake enters the food chain.

Complete the food chain.

Enter the initial for the animal that is both predator and prey in each food chain.

Balance of Predators and Prey

Predators and Prey help to keep ecosystems in balance.

If we removed all of the prey from an ecosystem then the predators in that ecosystem wouldn't have enough food and eventually they would stare.

Similarly if we removed all of predators from an ecosystem then the number of prey would increase.

The extra prey would be competing for a limited amount of food. Eventually the food would run out and the prey would starve and die.

Thats why having the right number of predators and prey; having predators and prey in balance, is vital to the health of all living things within an ecosystem.

Name: _____

Predators and Prey Quiz
Fill in the blank

Predators are animals that _____ other animals.

Prey are animals that are eaten by _____.

For example, an owl is a predator of a _____.

Some animals are _____ predators and prey.

For example, a robin might eat a _____, and

that same robin might get eaten by a _____.

Predators and prey keep ecosystems in _____.

They each have an important role. If there were no

_____, all the predators would starve to death.

If there were no predators, then the _____ of

the prey would increase so much that they would also

run out of _____.

food   population   balance   eat   both

hawk   grasshopper   predators   mouse   prey

Extinct and Endangered Animals

Complete this phrase.

Dead as a _____.

daffodil   dodo   doormouse   duck

Dead as a dodo means that something is definitely dead. The reason that we have this phrase is because the dodo was a flightless bird that was hunted to extinction about 300 years ago. When a species is extinct, it means that the very last member of that species has died.

Extinct and Endangered Animals

Dinosaurs roamed the earth for almost 200 million years. But about 65 million years ago, they suddenly became extinct.

Scientists have a few different ideas about why this happened. The most widely held theory is that a large rock from outer space, called an asteroid, smashed into earth.

This caused a heavy blanket of dust that blocked out the sun's light and caused a massive extinction of the earth's animals (like the dinosaurs and wooly mammoth) and plants.

More recently species have become extinct or are in danger of becoming extinct for a number of other reasons.

The dodo bird became extinct due to over hunting.

The Bali Tiger had become extinct due to hunting and the loss of its habitat. That's where the animal lives.

Pollution in the Yangtze river is thought to be the main reason that the Baiji Dolphin became extinct.

Some species such as the polar bear are threatened with extinction due to the effects of global warning which has reduced the number of days in the year in which there is sea ice that the polar bear needs in order to hunt seals.

Disease and viruses can also threaten animals with extinction. For example a contagious cancer has brought the tasmanian devil the brink of extinction.

Extinct species are ones that have died out forever and **endangered** species are ones that have come close to being extinct.

Which animals are extinct and which are endangered.

Draw a line through the extinct species and a box around the endangered species.

extinct

endangered

Dodo Bird

Bali Tiger

Woolly Mammoth

Polar Bear

Chimpanzee

Orinoco Crocodile

Baiji Dolphin

Giant Panda

Why am I extinct or endangered?

Fill in the blank under each animal that explains the likely reason the animal is extinct or endangered.

climate change   pollution   loss of habitat

over-hunting   asteroid?   disease

Connect the picture of the animal with its label and habitat.

What happens when a species becomes extinct?

Look at the food web below. In each of the two examples an animal has been marked as extinct. Explain what impact their extinction would have on the rest of the food web.

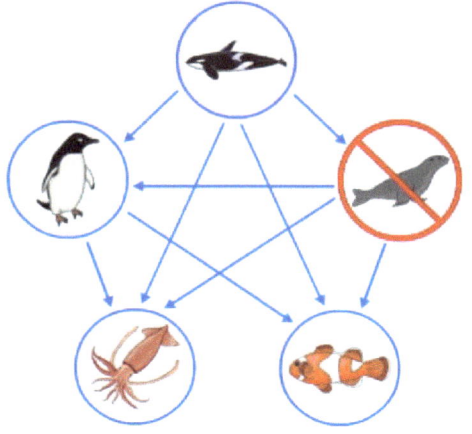

The Seal _____

_____

_____

_____

_____

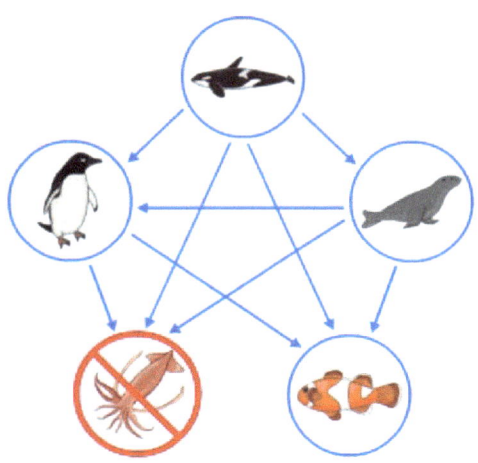

The Squid _____

_____

_____

_____

_____

How would the extinction of species affect the other?

What would happen to the species on the right if the species on the left became extinct. Use the arrows to indicate if the population would go up, down, or stay the same.

Name: _____

Extinct and Endangered Animals Quiz

1. Dead as a _____ means tha something is definitely dead.
   - a. Duck
   - b. Daffodil
   - c. Dodo
   - d. Doormouse

2. A species is said to be extinct…
   - a. When its very last member has died
   - b. When its last member is still alive

3. _____ species are ones that are close to becoming extinct
   - a. Endangered
   - b. Indanger

4. A polar bear is an extinct animal. True or false?

5. Dinosaurs became extinct of:
   - a. Disease
   - b. Pollution
   - c. Asteroid
   - d. Overhunting

6. The extinction of one species can threaten the survival of other species. True or false?

Newburyport, MA 01950

1-800-596-3175

OnBoard Academics employs teachers to make lessons for teachers! We create and publish a wide range of aligned lessons in math, science and ELA for use on most EdTech devices including whiteboard, tablets, computers and pdfs for printing.

All of our lessons are aligned to the common core, the Next Generation Science Standards and all state standards.

If you like our products please visit our website for information on individual lessons, teachers licenses, building licenses, district licenses and subscriptions.

Thank you for using OnBoard Academic products.